Paulo Tebulo
Sales Paiva
Banu Irenio

Evaluation of Atmospheric Nitrogen Fixation in the soil in Fabaceae

Paulo Tebulo
Sales Paiva
Banu Irenio

Evaluation of Atmospheric Nitrogen Fixation in the soil in Fabaceae

On the Agro-ecological Conditions of the District of Montepuez (Mapupulo) Province of Cabo-Delgado

ScienciaScripts

Imprint

Any brand names and product names mentioned in this book are subject to trademark, brand or patent protection and are trademarks or registered trademarks of their respective holders. The use of brand names, product names, common names, trade names, product descriptions etc. even without a particular marking in this work is in no way to be construed to mean that such names may be regarded as unrestricted in respect of trademark and brand protection legislation and could thus be used by anyone.

Cover image: www.ingimage.com

This book is a translation from the original published under ISBN 978-620-6-76054-2.

Publisher:
Sciencia Scripts
is a trademark of
Dodo Books Indian Ocean Ltd. and OmniScriptum S.R.L publishing group

120 High Road, East Finchley, London, N2 9ED, United Kingdom
Str. Armeneasca 28/1, office 1, Chisinau MD-2012, Republic of Moldova, Europe
Printed at: see last page
ISBN: 978-620-7-88741-5

Assessment of Soil Atmospheric Nitrogen Fixation in Fabaceae in the Agro-ecological Conditions of Montepuez District (Mapupulo) Cabo- Delgado Province.

Evaluation of Soil Atmospheric Nitrogen Fixation in Fabaceae in the Agro-ecological Conditions of the Montepuez District (Mapupulo) Cabo-Delgado Province.

Paulo Xavier Tebulo[1]

Faculty of Agronomic Sciences-Cuamba

Ptebulo@ucm.ac.mz

; Sales Alfredo Paiva[1]

Mandimba District Services for Economic Activities

Salespaiva3@gmail.com

Banu Belmiro Irenio[2]

Instituto de Investigação Agraria de Moçambique-Montepuez banumanhula@gmail.com

SUMMARY

The family sector in Mozambique practices traditional agriculture, which is characterized by accelerating soil degradation. In this way, an experiment was conducted in the experimental fields of the Agricultural Research Institute of Mozambique (IIAM) at the Mapupulo Agricultural Research Center (CIAM), with the aim of evaluating atmospheric nitrogen fixation in the soil in Fabaceae species, namely Soybean (*Glycine max* (L), Crotalaria (Crotalaria juncea (L), Nhemba Bean (*Vigna unguiculata (L)*, Boer Bean (*Cajanus cajans (L)* and Vulgar Bean (*Phaseolus vulgaris (L)*. DBCC was used, consisting of 5 blocks, the study variables were Number of Nodules per Plant (NNP), Weight of Nodules per Plant (PNP), Nitrogen Before Sowing (NAS), Nitrogen After Harvest (NDC). These variables were processed and analyzed using Analysis of Variance and Tukey's test at 5% significance in the R version 3.4.3 Patched package (2018-01-15 r74124). There were significant differences in all the variables studied. The species *Glycine max* had the highest number of nodules with 24.12±4.820, the lowest average was Phaseolus vulgaris with 11.40±1.738, in the middle were *Crotalaria juncea* with 16.74±3.610, Vigna unguiculata with 19.72±3.758 and Cajanus cajan with 11.68±2.922. The highest nodule weight was *Glycine max* with 2.688±0.312g/p, the lowest average was Cajanus cajan with 0.045±0.005g/p, in the middle were *Vigna unguiculata with* 1.686±0.212g/p, Phaseolus vulgares with 0.118±0.030g/p and *Crotalaria juncea* with 0.063±3.610g/p. N2 after harvest, the highest average was 450±6.124 N/kg/ha, with the lowest fixation rate was the species *Phaseolus vulgaris* with 292.4±112.611 N/kg/ha, in the middle were the species *Crotalaria juncea*, *Vigna unguiculata* and *Cajanus cajans* with averages of 423±17.176 N/kg/ha, 417±12.042 and 409.0±12.449 N/kg/ha. It is recommended that producers stick to the species with the highest fixation rate due to the results achieved, which is *Glycine max* with 450±6.124 N/kg/ha.

Key words: Legumes, nodules, atmospheric N2 fixation.

LIST OF ABBREVIATIONS

Abbreviation	Meaning
ANOVA	Analysis of Variance
SDAE	Economic Activities Services
CV	Coefficient of Variation
CIAM	Mapupulo Agricultural Research Center
DBCC	Causalized Complete Block Design
NAS	Nitrogen Before Sowing
NDC	Nitrogen after harvest
NNP	Number of nodules per plant
PNP	Weight of nodules per plant
NFC	Nitrogen fixed by the crop
Ha	Hectare
IIAM	Agricultural Research Institute of Mozambique
Kg	Kilogram
N	Nitrogen
MAE	Ministry of State Administration
Nm	Nanometer
BFNNL	Nodulating nitrogen-fixing bacteria in legumes
DARN	Directorate of Agronomy and Natural Resources
PH	Hydrogen Potential
g/p	Grams per plant

CHAPTER 1. INTRODUCTION

1.0 GENERALITIES

According to Rebah et al (2006) cited by Infante (2008), the fixation of atmospheric nitrogen in the soil represents around 70 to 85% of the total nitrogen used by plants. Around 80 million tons of nitrogen are fixed worldwide each year, compared to around 82 million tons obtained through industrial fixation.

According to Castro et al. (2004) leguminous plants have high levels of N in their roots, nodules and tissues during the flowering period, which means that they contribute more than 150 kg/ha/year of N, with between 60% and 80% of the N coming from biological nitrogen fixation.

Atmospheric nitrogen fixation is estimated to contribute around 258 million tons of N per year to different ecosystems, with agriculture contributing an estimated 60 million tons (Vieira, 2017).

Crotalaria (Crotalaria *juncea* L) is a tropical legume from the Fabaceae family, which comprises more than 500 species, Crotalaria juncea being one of them. The fixation of atmospheric nitrogen in the soil provides chemical, physical and biological improvements (Silva et al., 2014).

Crotalaria (*Crotalaria juncea* L) originated in India in Tropical Asia and is a legume with many characteristics, being an atmospheric nitrogen fixer (Arruda, 2012).

Legumes establish a symbiotic association with bacteria called rhizobia, which are responsible for the process of biological nitrogen fixation by communicating between the bacteria and the plant in the nodules, whereby the number of nodules and biomass production are intrinsically related (Silva et al., 2014).

Thus, the process of fixing atmospheric nitrogen in the soil is a viable alternative technique, both economically and ecologically, as it increases biomass production and consequently makes nitrogen available in the soil for growing subsequent crops (Silva et al., 2014).

The origin of the common bean crop (Phaseolus vulgaris L) and its diversification occurred in the Americas, but the exact location is still in doubt among authors. There is no consensus on this origin (Araujo, 2014).

There is, however, common sense among researchers today that the origin and domestication actually took place more than 7000 years ago in two centers of origin: Mesoamerica, Mexico, Central America and the Andean Region (Araujo, 2014).

Vulgar beans in their life cycle need high levels of nitrogen to achieve the highest desired yield, and the use of symbiotic bacteria contributes significantly to their yield, which ranges from 1.8 to 3.5t/ha (Oliveira, 2016).

The bean root system is upright and shallow and has nodules on the lateral roots due to symbiosis with rhizobium (Atmospheric Nitrogen-Fixing Bacteria) (Ferreira, 2015).

The nhemba bean (*Vigna unguiculata* L. Walp) originated in Africa, where it spread to India, China, Central America and North America. Its expansion in Africa and the rest of the world was thanks to migration, trade contacts and wars (Infante, 2008).

The nhemba bean is a dicot of the Leguminosae family, Subfamily papilionidade, genus Vigna, species Vigna unguiculata L. It is a species with a high level of atmospheric nitrogen fixation (Infante, 2008).

The soybean crop (Glycine max (L.) Merrill) originated in China, in Manchuria, the central region of China, where it appeared approximately 5,000 years ago and was spread all over the world by Chinese, Japanese and English travelers and emigrants (Mendes, 2019).

Soybeans belong to the kingdom Plantae, family fabaceae, genus glycin, species Glycin max L (Mendes, 2019).

The use of crops from the fabaceae family to enrich the soil is a practice that has been known for more than two millennia by the Chinese, and then used by the Greeks and Romans (Araujo at al, 2019).

1.1. Research problem

Chemical fertilizers are expensive, the biological process with bacteria belonging to the Bradyrhizobium genus can supply all the N needed by crops, coupled with the limitation of the nitrogen reservoirs present in the soil's organic matter, which can be depleted quickly after a few crops, the temperature and humidity conditions caused by global warming accelerate the decomposition of organic matter and N losses, resulting in soils with poor N contents, capable of leveraging crops in the family sector (Silva et al., 2014).

Nitrogen fertilizers are the form of fertilizer most readily assimilated by plants, but they are expensive and beyond the capacity of family producers, and their productivity boosts food security in our country (Hungria et al, 2001).

The steps that transform N2 into (NH3) require hydrogen derived from petroleum gas, an iron-containing catalyst, high temperatures (300 to 600°C), high pressures, and gas from non-renewable energy sources (Hungary et al, 2001).

The use of nitrogen fertilizers lies in the low efficiency of plant absorption, rarely exceeding 50%, i.e. if the farmer fertilizes 100kg of N2 in the soil, 50kg will be lost in a short space of time through the processes of washing into the soil profiles by percolation or surface runoff (leaching) and transformation into gaseous forms (Hungria et al, 2001).

1.2. Justification

The production of various crops in most tropical soils is limited by the need for nitrogen fertilization. Among the mineral macro-nutrients essential to plants, nitrogen is one of the most limiting for plant growth, although it accounts for more than 78% in atmospheric air (Silva, 2018).

Among the atmospheric nitrogen-fixing bacteria present in the soil, whose populations are highly influenced by the rhizosphere of plants, rhizobium is considered the most significant group in our country's agriculture for its association with legumes (Silva, 2018).

Growth time in the culture medium, for example, is an important characteristic for the preliminary characterization of nitrogen-fixing bacteria that nodulate legumes. Initially, fast-growing bacteria were considered to be those species that took 2 to 3 days to present isolated colonies in the culture medium (Brocardo, 2013).

Today, family farmers intercrop leguminous plants with different crops, but they don't know which species is the most fixative and efficient, given the poor soils and the low yields of agricultural products in the province of Cabo-Delgado, in the district of Montepuez, Mapupulo Administrative Post.

In order to respond to the problematization set out above, this work will study the evaluation of atmospheric nitrogen fixation in the soil in fabaceae species in the agro-ecological conditions of the district of Montepuez (Mapupulo) Province of Cabo-Delgado, in order to identify the species that fix the most nitrogen to later be indicated for intercropping, crop rotation and even control of degraded soils.

1.3. Objectives

1.3.1 General

> Evaluate the fixation of atmospheric nitrogen in the soil in fabaceae species.

1.3.2 Specific objectives

> Determine the number of nodules in each treatment

> Determine the weight of nodules in each treatment

> Determine the amount of nitrogen in the soil before sowing and after harvesting in all treatments.

1.4. Hypotheses

Ho: There are no significant differences when it comes to fixing atmospheric nitrogen in the soil.

Ha: At least one of the legumes will show significant differences when fixing atmospheric nitrogen in the soil

CHAPTER II. LITERATURE REVIEW

2.0 Legumes

2.1 Origin of the Crotalaria juncea crop

Crotalaria is a tropical legume from the Fabacea family, which includes more than 500 species, including *Crotalaria juncea*. The *Crotalaria juncea* crop originated in Asia, specifically India, and is capable of adapting to tropical climates (Silva et al., 2014).

2.2 Taxonomic classification of Crotalaria

Taxonomy: family Fabaceae, subfamily Faboideae, tribe Crotalarieae, genus Crotalaria, species Crotalaria *juncea* L. (Araujo, 2015)

2.2.1 . Characteristics of Crotalaria juncea

Crotalária juncea blooms in short days, with 10 to 20 seeds per pod and has a high nitrogen-fixing capacity in conditions without water stress and can grow rapidly in the first few days after emergence (Araújo *at al*, 2019).

This crop is grown in tropical and subtropical climates and can be grown in clay and sandy soils. It is very efficient at fixing atmospheric nitrogen (Lamônica, 2008).

Crotalaria juncea has the potential to fix atmospheric nitrogen in the soil at around 150 to 450 kg/ha/year compared to pigeonpea, which is around 37 to 280 kg/ha. *Crotalaria juncea* can reach a height of 2 to 3.5 m (Filho, 2015).

At 130 days after sowing, the roots can reach a depth of 4 or 5 m, depending on the soil, and under these conditions they produce a vegetable biomass that is used for green manure and ground cover due to these characteristics. (Filho, 2015)

Crotalaria juncea is a short-day flowering legume, flowering at 60 days with a 120-day cycle. The pods have 10 to 20 seeds, but the root cannot break through compact, waterlogged soils (Araújo, 2015).

2.2.2 Vulgar Culture

2.2.3 Origin

Vulgar bean cultivation originated in America, but many hypotheses maintain that the first place it was identified was in Mesoamerican Mexico and Guatemala, in the south of the Antes, from Peru to northeastern Argentina, in the north of the Antes, specifically in Colombia and Venezuela (Diniz, 2006).

2.2.4 Taxonomic classification

The common bean (Phaseolus vulgaris L.) belongs to the Plant Kingdom, Sub-branch Angiosperm, Class Dicotyledonous, Order Fabales, Family Fabaceae, Subfamily Faboideae, Tribe Phaseoleae, Genus Phaseolus, Species Phaseolus vulgares L. (Ferreira, 2015).

2.2.5 . Morphology of the Vulgar Bean crop.

2.3 Flower

The flowers of the bean plant are perfect, i.e. they have male and female organs in the same flower. They are also complete, with a corolla and calyx, the male organs are made up of 10 stamens, the corolla is made up of 5 petals joined at the base. (Oliveira, 2016)

2.3.1 Fruit

The fruit of the bean plant is a pod of various shapes depending on the variety, which can be straight, slightly curved, slightly rounded or flattened. The length of the pod can vary between 9 and 12 cm and has 307 seeds per pod (Ferreira, 2015).

2.3.2 Nitrogen fixation from nodules

The efficiency of symbiosis in nitrogen fixation depends on the size of the nodules, so the greater the nodule mass and the greater the nitrogen fixed. Various sizes of nodules have been classified, which can be large > 3.35 mm, medium between 2.00 and 3.35 mm, small between 2.00 and 1.00 mm and very small < 1.00 mm (Matoso, 2013).

According to Oliveira & Sbartelotto (2011), they observed that regardless of inoculation with rhizobium, the bean plant presents nodules indicating the presence of strains native to the soil. Good nodulation should be greater than 20 nodules per plant, which should be red in color as a sign of the life and activity of the bacteria and nitrogen.

In the cultivation of common beans, symbiosis occurs in the roots with the Rhizobium bacterium. When this bacterium is present in the soil, it naturally recognizes and infects the roots of the bean causing the formation of nodules, where the fixation of atmospheric nitrogen occurs (Araújo et al., 2007).

The biological N2 fixation of common beans (Phaseolus vulgaris L.) is not as efficient as other grain legumes, partly due to the lack of synchrony between the depletion of cotyledon reserves and the formation of nodules and effective N fixation, which causes nutritional stress at the beginning of the cycle (Brito, 2013).

2.3.3 Growing Boer Beans

2.3.4 Origin

The cowpea Cajanus cajan (L.) was domesticated thousands of years ago in India, and its cultivation has several advantages from an agronomic perspective. Cowpea plants have a positive effect on soil fertility with the potential to fix around 235 kg of atmospheric N2 per hectare (Oppewal & Cruz, 2017).

Generally speaking, the pigeonpea originated in India and was introduced to other continents on the slave route, including Africa, where it became a source of human food, as it is a plant of tropical or subtropical origin that grows well in the climates of Mozambique (Carellos, 2013).

2.3.5 Taxonomic classification

The cowpea is a legume from the kingdom plantae, division Magnoliophyta, class Magonoliopsida, order fabales, family Fabaceae, sub-family Faboideae, genus Cajanus, species Cajanus cajan (L) (Silveira, 2010).

2.3.6 Botanical characteristics

The pigeonpea is an erect, shrubby plant, some varieties can reach 4 m and others no more than 1 m. It has semi-woody stems and a pivoting root that can have fine, secondary roots 30 cm above the topsoil (Oppewal & Cruz, 2017).

Its roots have nodules containing symbiotic atmospheric nitrogen-fixing bacteria. The cowpea fixes nitrogen at between 120 and 350 km/há/year (Carellos, 2013).

The bean crop has 3 leaves, with lanceolate or elliptical leaflets 4 to 10 cm long. The flowers are in terminal racemes 1.5 to 1.8 cm long, yellow or orange-yellow or reddish in color, with green indehiscent pods (Carellos, 2013).

2.4 Boer bean morphology

The pigeonpea has a pivotal root with numerous fine secondary roots that can reach 30 cm into the soil and the main root can reach a depth of 1 m. They have nodules and bacteria of the Rhisobium genus that live symbiotically with the plant and help to fix atmospheric nitrogen, which is given to the plant to form amino acids (Seiffert & Thiago, 1983).

2.4.1 Leaf

The leaves of the pigeonpea crop are trifoliate, light green in color, with silky pubescence and a length of between 2.5 and 9 cm (Souza, 2007).

2.4.2 Flowers

According to Sousa (2007), Cajanus cajan flowers are hermaphroditic, with a very high self-pollination rate of around 40 to 70%. The inflorescence produces two to eight pods and is yellow or orange in color, with axillary racemes between 5 and 12 and a length of 1.5 to 1.8 cm.

For the nhemba bean, there is no conclusive information on the minimum number of nodules needed to guarantee good BNF performance, as is observed for the soybean crop, for which 15 to 20 nodules on the crown of the main root are recognized as sufficient (Hungria & Bohrer, 2000). Cited by (Melo & Zilli, 2009)

Some varieties of nhemba bean have over 40 nodules per plant, thus providing greater mass and more nitrogen fixed in the plants and soil for future crops or to improve degraded soils (Melo & Zilli, 2009).

2.4.3 Growing Nhemba beans

2.4.4 Origin

The nhemba bean (Vigna unguiculata L. Walp) originated in Africa, where it spread to India, China, Central America and North America. Its expansion in Africa and the rest of the world was thanks to migration, trade contacts and wars (Infante, 2008).

The nhemba bean crop improves soil fertility due to its ability to fix atmospheric nitrogen, helping to increase yields of maize and other nitrogen-demanding crops grown in intercropping or rotation systems (Alfredo, 2013).

2.4.5 Taxonomic classification

The nhemba bean is a dicot of the Leguminosae family, subfamily papilionidade, genus Vigna, species Vigna unguiculata L. It is a species with a high level of atmospheric nitrogen fixation (Infante, 2008).

The nhemba bean crop is resistant and tolerant to water conditions, and also has the ability to nodulate efficiently with bacteria of the rhizobium genus, thus reducing the cost of synthetic pesticides and fixing more than 100 kg of N/ha required in all phenological phases (Alfredo, 2013).

2.4.6 Soybean cultivation

2.5 Origin

The soybean crop (Glycine max (L.) Merrill) originated in China, in Manchuria, the central region of China, where it emerged approximately 5,000 years ago and was spread all over the world by Chinese, Japanese and English travelers and emigrants (Mendes, 2019).

2.5.1 Taxonomy

Soybeans belong to the kingdom Kingdom: Plantae Division: Magnoliophyta Class: Magnoliopsida Order: Fabales Family: Fabaceae (Leguminoseae) Subfamily: Faboideae (Papilionoideae) Genus: Glycine Species: Glycine max (L.) Merrill (Mendes, 2019)

Soybean cultivation helps biological nitrogen fixation through nodule activity, improving plant development and productivity (Hirakuri et al., 2014). Cited by (Oliveira et al., 2018)

An alternative for reducing the use of agrochemicals is the addition of organic fertilizers or biofertilizers, which incorporate organic matter and nutrients into the soil that are essential for plant development, thus helping to minimize environmental impact and soil health (Oliveira et al., 2018).

Soybeans are a crop with a high demand for nutrients, especially nitrogen, which is extracted predominantly through biological fixation by nitrogen-fixing bacteria, which translates around 150 kg ha-1 of nitrogen into the soybeans and into the soil (Brandelero, 2009).

2.5.2 Soybean crop morphology

2.5.3 Root

The pivoting soybean root with the main taproot with a large number of secondary roots (diffuse), which can reach a length of 1.8 m (usually 0.15 m) - nodules are present (Embrapa, 2011).

2.5.4 Leaves

The leaves are simple embryonic or unifoliate and can be compound or trifoliolate, varying in green color and shape. (Martins, 2016)

2.5.5 Flowers

The flowers of the soybean plant are complete (calyx, corolla, androecium and gynoecium) that appear in terminal or axillary racemes, with white or purple coloration characterized by flowering induced by photoperiodism for each inflorescence reaches 8 to 40 flowers (Martins, 2016).

As an efficient fixer of atmospheric nitrogen, the soybean crop can fix 100 to 160 kg of N2 per hectare, which can benefit other successor crops, but for this to happen effectively, it is important that the temperature is suitable for nitrogenase activity (Pereira, at al., 2011).

The formation of nodules on the roots is the result of a complex process, involving several phases that begin with the germinating seed and the roots exuding molecules that chemically attract the rhizobia, others that stimulate the growth of the bacteria, then the bacteria penetrate the soybean roots and cause the growth of specific cells (Hungria *et al*, 2001).

For nodulation to occur, some factors are decisive in the biological fixation of N2 by legumes, with water tension, O2 content in the nodule, soil temperature and pH, salinity, toxins and predators being the main ones that can act together with the wide variety of rhizobium strains found in the soil (Souza, 2015).

The vascular system develops positive turgor in response to the transport of nitrogen compounds via mass flow. This assumes that export requires water and that water is indispensable in this flow, since the water that brings sucrose via the phloem is absorbed by the nodule on its way back, carrying the nitrogenous solutes (Souza, 2015).

The formation of nodules involves a series of processes involving signaling molecules between the plant and the bacteria. The host plant secretes substances such as amino acids, CO2, hormones, organic acids, sugars, vitamins, polysaccharides, proteins and flavonoids (Santos, 2013).

Nodulation occurs approximately 2 hours after the bacteria come into contact with the roots. The formation of nodules on the roots is the result of a complex process, involving several phases that begins with the germinating seeds and the roots exuding molecules that chemically attract the rhizobia, others that stimulate the growth of the bacteria. (Hungria et al, 2001).

For nodulation to occur, some factors are decisive in the biological fixation of N_2 by legumes, with water tension, O_2 content in the nodule, soil temperature and pH, salinity, toxins and predators being the main ones that can act on the wide variety of rhizobium strains found in the soil (Souza, 2015).

The vascular system develops positive turgor in response to the transport of nitrogen compounds via mass flow. This assumes that export requires water and that water is indispensable in this flow, since the water that brings sucrose via the phloem is absorbed by the nodule on its way back, carrying the nitrogenous solutes (Souza, 2015).

The formation of nodules involves a series of processes involving signaling molecules between the plant and the bacteria. The host plant secretes substances such as amino acids, CO_2 , hormones, organic acids, sugars, vitamins, polysaccharides, proteins and flavonoids (Santos, 2013).

Nodulation occurs approximately 2 hours after the bacteria come into contact with the roots. Primary nodules develop in regions of elongation and in areas where small root hairs are formed, which is considered to be the preferential region for infection by the fixing bacteria (Santos, 2013).

develop in regions of elongation and in the areas where small root hairs are formed, which is considered to be the preferential region for infection by the fixing bacteria (Santos, 2013).

2.5.6 Number of nodules per plant in the crotalaria crop

The number of nodules on the *crotalaria juncea* plant depends on the amount of atmospheric nitrogen-fixing bacteria and or plants inoculated with these fixers. According to Silva et al. (2012), plants reach 50 nodules in ferralitic soils in pots.

2.6 Number of nodules per plant in the nhemba bean crop

The nhemba bean crop shows nodulation, with an association with nitrogen-fixing bacteria (BFN), and 345 nodules were obtained from the soil samples collected, which found an average of 53.57 nodules per soil sample, and an average of 14.42 nodules per cultivated plant (Souza, et al, 2013).

2.6.1 Number of nodules per plant in the soybean crop

The soybean crop has an average of 30 nodules per plant in a natural environment, thus helping root development, which will increase nodulation and, consequently, BNF, improving plant growth and development (Oliveira et al, 2018).

The number of nodules has a direct correlation with nitrogen fixation capacity (BNF) in some nitrogen-fixing crops, where a plant with good nodulation will absorb more N, reducing the use of external sources of N, such as chemical fertilizers (Oliveira *et al*, 2018).

2.6.2 Number of nodules in the common bean crop

In the common bean crop, nodules are formed in the range of 3 to 20 nodules on each plant, approximately 4 to 5 mm in diameter and with a reddish color inside as an indicator of the activity of fixing N in the soil to feed the plants (Oliveira & Sbardelotto, 2011).

2.6.3 Number of nodules on cowpeas

In cowpea cultivation, the number of nodules per plant varies between an average of 3.5 and 30.5. These results depend on the type of variety and the bacteria present in the soil capable of infecting the roots and forming nodules (Oliveira, 2009).

2.6.4 Fixing atmospheric nitrogen

Atmospheric nitrogen fixation is an enzymatic process in which N_2 is reduced to NH3 by the action of free-living microorganisms associated with host plants in symbiosis; these organisms are diazotrophs, i.e. with the ability to fix N2. (Vieira, 2017)

N2 fixation occurs in structures called nodules. After the nodules are formed, various proteins are synthesized and one of them, leghemoglobin, controls the supply of oxygen to the nodule tissues (Hungria *et al*, 2001).

N2-fixing bacteria can be divided into three, non-symbiotic or free-living fixers, associative fixers, which form a casual relationship, and symbiotic fixers that fix N_2 in organized associations with higher plants (Vieira, 2017).

2.6.5 Table 1: Examples of nodulation gene inducers and host legumes

Legumes	Rhizobium species
Glicine max	Bradyrhizobium j aponicum
Phaseolus vulgaris	Rhizobium tropici
Crotalaria juncea	Methylobacterium nodulans
Cajanus cajan	Methylobacterium ssp
Vigna unguiculata	Bradyrhizobium sp

Source: (Vieira, 2017)

2.6.6 Fixation of atmospheric N2 in the soil

According to Castro *et al.* (2004). The *crotalaria juncea* crop fixes around 34.4 kg/ha, totaling 85.9% fixation.

Table 2 of nitrogen fixation levels in some legumes

Scientific name	Common name	Amount of N fixed Kg/Ha
Cajanus cajan	Guandu	37-280
Canavalia ensiformis	Pork beans	49-190
Crotalaria brevifora	Crotalaria	98-160
Crotalaria juncea	Crotalaria	150-450
Crotalaria mucronata	Crotalaria	80-160
Crotalaria ochroleuca	Crotalaria	133-200
Crotalaria spectabilis	Crotalaria	60-120
Dolichos lab-lab	Labelabe	66-180
Lathyrus sativus	chicharo	80-100
Mucuna aterrima	Black mucuna	128-268
Mucuna cinereum	Mucuna gray	170-210
Mucuna deeringiana	Mucuna ana	50-100
Vicia sativa	Vetch	90-180

Source: Derpsch & Calegari (199); Wutke (1993)

2.7 Atmospheric nitrogen fixation in the Boer Bean crop.

In soils with low artificial nitrogen content, the highest bean yields are observed under Rhizobium inoculation, culminating in the fixation of more than 90% of atmospheric N2, which is effectively equivalent to 144 to 199 kg/ha of N_2 depending on the type of Rhizobium (Rufini, 2013).

This symbiosis takes place in the nodules located on the roots. There is a harmonious exchange relationship in which the plant supplies energy to the bacteria and the bacteria receive ammonia produced by *Rhizobium from the* fixation of atmospheric nitrogen (Souza *et al*, 2007).

Ammonia is translocated from the nodules and distributed throughout the plant, being located in greater volume in the leaves, young tissues and seeds, where it participates in the formation of amino acids and proteins (Silveira, 2010).

It is estimated that biologically fixed nitrogen in the boer bean crop can reach up to around 200 kg/ha per year depending on soil conditions, compaction and cation exchange capacity (Souza *et al*, 2007).

2.7.1 Table 3: Atmospheric N2 Fixation Levels

Species	Rates of biological N2 fixation in the soil kg/ha/year
Vigna unguiculata	**73-240**
Cajanus cajans	**7-235**
Glycine max	**17-450**

(Simione *at al.*, 2014)

Table 4: Examples of some nodulating legumes with respective rates of biological N2 fixation in the soil

LEGUMINOUS - Scientific name	KG/HA/YEAR
Alfalfa (*Medicago sativa*)	127-333
Calopogonium (Calopogonium mucunoides)	64-450
Groundnuts (*Arachis hypogaea*)	33-297
Caupi *(Vigna unguiculata)*	73-240
Crotalaria (*Crotalaria juncea*)	197-249
Stylosanthe (Stylosanthe ssp.)	20-263
Gunado (*Cajanus cajan*)	7-235
Leucaena (Leucaena leucocephala)	400-900
Soybeans (*Glycine max*)	17-450

Source: Moreira & Sequeira cited by Calengari *et.al* 1993 ; Hardarson (1993); People *et.al* (1995)

CHAPTER III. MATERIAL AND METHODS

3. Description of the study site

Montepuez District is located in the southern part of Cabo- Delgado Province, 210km from the provincial capital - Pemba. It borders Mueda District to the north, Namuno and Chiure Districts to the south, Ancuabe and Meluco Districts to the east and Balama and Mecula Districts to the west, the latter being in Niassa Province.

With a surface area of 17,721 km2 and a population of 149,181 inhabitants recorded in 1997 and estimated at 186,476 inhabitants as of 1/1/2005, the Montepuez district has a population density of 10.5 inhabitants/km2.

Climatically, the Montepuez district is dominated by a semi-arid, dry sub-humid climate. The average annual rainfall varies between 800 and 1200 mm, while the reference potential evapotranspiration (ETO) is between 1300 and 1500 mm, making it a rainy sub-humid climate.

In terms of the average temperature during the growing season, there are regions where temperatures exceed 20 and 25° C. The district is crossed by important non-permanent rivers throughout the year, with the exception of the Lugenda River, which serves as a boundary with the Mecula district in neighboring Niassa Province.

3.1 Study area

This study was carried out in the experimental field of the Centro de Investigación Agraria de Mapupulo (CIAM), in the legume sector from January to June

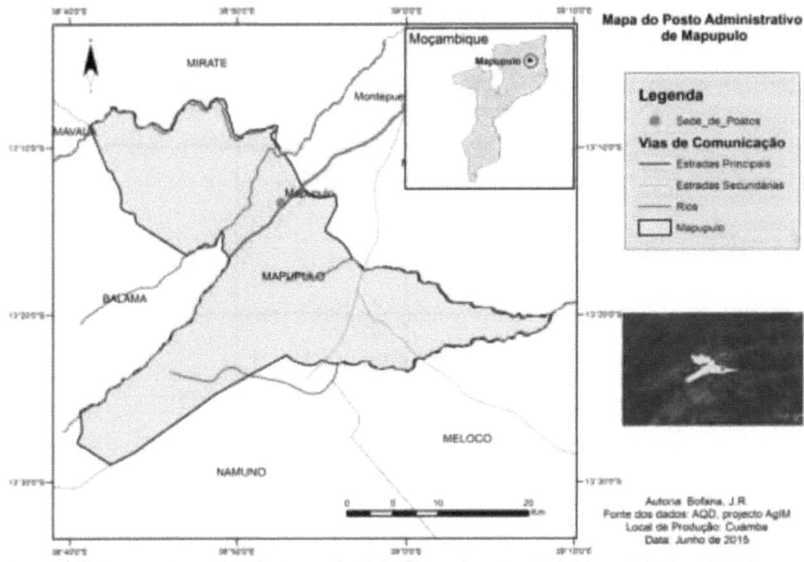

Figure-1 Geographical location of the Mapupulo administrative post.

3.2 Description of the study area

The Mapupulo Agricultural Research Center (CIAM) is located in the Mapupulo Administrative Post, 18 km from the district town of Montepuez, at the entrance to the Balama district. It has a total area of 55 hectares, of which 20 hectares are used for research programs and another 35 hectares for the production of various crops.

3.2.1 Total nitrogen analysis procedures

3.2.2 Reagents

To analyze total nitrogen you need some reagents after laboratory processes, some reagents are recommended such as: concentrated sulphuric acid, hydrogen peroxide (30%) and sodium hydroxide (35%).

350g of sodium hydroxide was placed in a 1L triangular flask, 800ml of distilled water was added and dissolved using a magnetic stirrer. The entire solution was poured into the 1L triangular flask. A funnel was used in a 1L volumetric flask. The 1L triangular flask was rinsed 3 times with distilled water so that the solution contained in the 1L volumetric flask formed a volume of 1L.

3.2.3 Hydrochloric acid (0.05N)

0.42 ml of concentrated hydrochloric acid is poured into a 100 ml volumetric flask and distilled water is added to make a volume of 100 ml.

3.2.4 Boric acid 4%

40 mg of boric acid was placed in the 1L triangular flask, 500 ml of distilled water was added and dissolved by heating, and the concentration of the solution was adjusted to 5.0 Ph. Using a funnel, the entire solution contained in the 1L triangular flask was poured into the 1L volumetric flask. The 1L triangular flask was washed three times. With distilled water, all the washing liquid was poured into the 1L volumetric flask. After adjusting the samples, they were placed in a colored glass and kept in a cool, dark place.

3.2.5 Analysis procedures for samples preserved in colored glass

Decomposition

0.5 g of soil was weighed, placed in the decomposition glass and 4 ml of concentrated sulphuric acid and 2 ml of hydrogen peroxide were added. After a violent reaction, a microwave digester was used and heated. It was left to rest until the temperature returned to normal.

3.2.6 Quantitative nitrogen analysis

All the decomposition liquid was transferred and added to the test tube using 50 ml of deionized water, then distilled and titrated using an automatic distiller. After this step, a blank test was carried out, which consisted of a sample without soil in a test tube with water and reagent as recommended.

3.3 Quantitative analysis of NO3-N, 5% salicylic acid solution

1 g of salicylic acid was placed in a 50 ml tube and dissolved by adding 20 ml of concerted sulluric acid. This solution was kept in the refrigerator and used for two weeks.

3.3.1 2N sodium hydroxide as one of the reagents for quantities of N

80g of sodium hydroxide was weighed and placed in a 500 ml beaker, 400 ml of distilled water was added and dissolved using a magnetic stirrer. Using a funnel, the entire solution from the 500 ml beaker was poured into a 1L volumetric flask. The 500 ml beaker was washed three times with distilled water and all the washing liquid was poured into the 1 L volumetric flask. With distilled water, the solution contained in the 1L volumetric flask was made up into a volume of 1L. Formed a volume of 1l.

3.3.2 Sample analytical curve

The 1000 ppm NO3-N solution purchased commercially was diluted with 1N KCL to create solutions of 0;2;5;5.7;5;10 ppm concentration.

3.3.3 Analysis procedures

Extraction as one of the steps in soil analytical curve analysis A quantity of soil was extracted, weighed and 2 g of soil was placed in a 50 ml container for the typing apparatus. Once the stirring process was complete, 20 ml of 1N KCL was added and stirred for 30 minutes, then the extracted liquid was filtered through a paper filter.

3.3.4 Quantitative analysis of NH4-N

0.2 g of heavy magnesium oxide was added to 10 ml of extracted liquid and steam distilled.

In a 250 ml triangular flask with 50 ml of 2% boric acid, distil until 100 to 150 ml of distilled liquid has accumulated and titrate with 0.01N hydrochloric acid until the solution turns pink.

3.3.5 Quantitative analysis of NO3-N

1ml of the filtered liquid was placed in a 15 ml tube, heated to a temperature of 150° C using an oven and dried and hardened. 0.4 ml of 5% salicylic acid solution was added and left to stand for 5 minutes. 10 ml of 2N sodium hydroxide solution was added, mixed well and left to rest until it returned to normal temperature. A spectrophotometer was used to measure the degree of absorbance of the waves at a length of 410 nm.

3.4 Experimental material

For this study, 5 leguminous crops were used as experimental material: *Crotalária juncea*, Feijão vulgar, Feijão Bóer, Feijão nhemba and soybeans.

Table 5: Description of the treatments under study

Treatments	Cultures	Variety
T1	**Crotalaria juncea**	**Crotalaria**
T2	**Boer beans**	**ICAP00557**
T3	**Soy**	**Sun babi**
T4	**Kidney beans**	**White gold**
T5	Nhemba beans	IT13

Source: The authors of the study

3.4.1 Experimental design

For this study, a causalized complete block design (DBCC) with 5 blocks was used, with a mono-factorial test scheme with 5m long plots corresponding to 25 plots in a total area of 696m2.

3.4.2 Experimental conduct

3.4.3 Soil preparation

The soil was prepared mechanically in November using a tractor coupled with a disc plough. The plowing depth was 25cm, and with the help of long-handled hoes it was possible to level it out, with the aim of fluffing the soil so that it would receive the seed without clods, with a porosity and moisture retention capacity.

3.4.4 Taking the First Soil Sample

After plowing, the initial soil sample was taken in order to check the level of nitrogen concentration in the soil before sowing the seeds. This process was carried out using a diagonal technique that included 9 points marked on each plot using two ropes, a ruler and a tape measure.

Five samples were taken from each block, in five study treatments, plot by plot, mixing the samples from the same treatments. At each point, 3 samples were taken vertically at a depth of 30 cm, first at points 0 to 10, 10 to 20 cm, and 20 to 30 cm. The total amount of a sample for each treatment was 2 kg of soil, and it was stored at room temperature to be sent to the laboratory.

3.4.5 Sowing

Sowing was done by hand on February 14, 2020, using different spacing between plants, the spacing between plants was the same for all treatments: T1=Crotalaria,70cm*10cm, T2=Boer bean,70cm*30cm, T3=Soybean,70*10, T4=Vulgar bean,70cm*15cm, T5=Nhemba bean, 70cm* 15cm, with a depth in the range of 2 to 3cm. The difference in caliper in the same trial is because the trial supports different crops.

3.4.6 Emergency

The first emergence was seen in the crotalaria crop at 4 days, with more than 25% of the seeds having emerged by 8 days, but the pigeonpea treatment was the most delayed due to the depth of the sowing. Some treatments, such as common bean and pigeonpea, had to be sown in a few pits where there was no emergence due to the soil, which is characterized by greater plasticity.

3.5 Sachets

During the trial, three (3) weeding sessions were carried out, all treatments, the first weeding was on March 4, 2020 and the second weeding was on March 15 and the last weeding in the second week of April of the year written above.

3.5.1 Phytosanitary control

Phytosanitary control was focused on the two groups of chewing and sucking pests in all treatments, such as the bean fly, *Ootheca mutabilis,* bedbugs and the leaf caterpillar (*Lamprosema indica*). The first spraying took place on February 26th, using cypermethrin 2.5% EC at a dose of 0.5ml/L of water.

3.5.2 Taking the second soil sample

This process was carried out after the crops had reached 75% flowering. Soil samples were taken from each treatment using a diagonal technique, and the sample taken was placed in a sealed plastic bag and kept at room temperature.

3.5.3 Taking the sample after harvesting the crops

For this variable, 5 samples were taken from each block in all the treatments after the legumes had been harvested, using two ropes, hoes, a knife and a 30 cm ruler. A sample was taken from each treatment after the harvest to measure the rate of fixation of each legume, using a few marked points at which the sample was taken at a depth of 30 cm and mixed by treatment into a sample weighing 2 kg.

3.5.4 Harvesting

Harvesting was carried out manually when the plants reached 75% flowering, after which nodules were counted and weighed, and a soil sample was taken after sowing.

3.5.5 Measurement variables

The variables measured and observed were: number of nodules per plant, weight of nodules per plant, soil sample taken before sowing, analysis of soil taken after 75% flowering and nitrogen after harvest.

3.6 Soil sampling Before sowing

The soil was collected before sowing after the trial scheme was sketched out on paper and then the scheme was drawn up in the field to make the process easier, and soil samples were taken from each plot. The soil samples were taken using the diagonal technique, at a depth of 30 cm, with a total weight of 2 kg of soil for each sample.

3.6.1 Number of nodules per plant

For this variable, it was done after the crops had reached 75% flowering, by randomly selecting 20 plants per crop, using a hoe along with a 10-liter bucket of water to wash the root part of the harvested plants and then counting the nodules on each plant.

3.6.2 Weight of nodules per plant

In this variable, after the legumes had reached 75% flowering, 20 plants were randomly selected for each crop, the root part was washed, then the nodules were extracted with the help of a knife, the larger nodules could be harvested by hand and weighed using a HR-200 precision scale.

3.6.3 Data processing

The data obtained was entered and organized in a Microsoft Excel spreadsheet and processed in the R package version 3.4.3 Patched (2018-01-15 r74124) for analysis of variance (ANOVA). This first test helped to process and interpret the raw data obtained in the experiment to determine whether or not there was a significant difference between the treatments. Tukey's test at a 5% significance level was then used to compare the means.

Table 6. Single-factor anova scheme for DBCC

Diagram of the Analysis of Variance for this study

Causes/Sources of Variation	GL	Sum of Squares	Squares Media	F Calculated
Blocks	J-1	SQBlocks	QMBlocks	QMBlocks/QMRes
Treatments	1-1	SQTrat	QMTrat	QMTrat/QMRes
Waste	(I-1)(J-1)	SQRes	QMRes	
Total	IJ-1	SQTotal		

Source: Anjos, 2005

The statistical model of the randomized block design is given by:

$$ij\ i\ j\ ij\ Y = p + a + b + e$$

in which:

$ij\ y$ is the value observed in the experimental plot that received treatment i in the block

j;

LI represents a general constant associated with this random variable;

j bis the effect of block j (j =1, 2,...,b);

i a

is the effect of treatment i (i =1,2,...,t);

ij eis the experimental error.

3.6.4 Coefficient of Variation

The coefficient of variation is a measure of dispersion used to estimate the precision of experiments and represents the standard deviation expressed as a percentage of the measurement. As a measure of dispersion, the main quality of the coefficient of variation is its ability to compare the results of different studies involving the same response variable, allowing the precision of research to be quantified (Padovani, 2014).

3.6.5 Constraints

After sowing, the crops did not fully emerge, due to the plasticity of the soil and post-emergence insect attacks. The search for capital, which took time to pay for the samples in the laboratory, was a major constraint in this research.

CHAPTER IV. RESULTS AND DISCUSSION

4.1 Number of nodules per plant

Table 7 shows the average results for the variable number of nodules per plant harvested from different treatments and species.

Table 7: Results for the number of nodules per plant (NNP)

Treatments	Mediai Standard Error
Crotalaria	$16.74i3.610^{bc}$
Boer beans	$11.68i2.922^{c}$
Nhemba beans	$19.72i3.758^{ab}$
Kidney Beans	$11.40i1.738^{c}$
Soy	$24.12i4.820^{a}$
CV (%)	21.036
General Media	16.732

Means followed by the same letters in the columns in the table do not differ when Tukey's test is applied at the 5% error level.

According to the results of the analysis of variance at 5% significance, the species showed significant differences for the variable number of nodules per plant. Table 7 shows the averages for the number of nodules per plant for the different species. There was a variation between the maximum average number in the soybean crop with 24.12 followed by the nhemba bean crop with 19.72 and Crotalaria with 16.74, minimum in the cowpea was 11.68 and 11.40 for common bean.

Similar results to this study were obtained by Brandelero *et al.* (2009) in their study of 9 soybean varieties in terms of nodulation, it was found that the variety with the highest number of nodules per plant reached an average of 23.4.

According to Infante (2008), in his study analyzing the formation of root nodules and nitrogen fixation in 6 varieties of nhemba bean (*Vigna unguiculata*), the results with the lowest average were the Tete-2 variety with 18.2 and the IT82D-812 variety with an average of 39 nodules per plant.

According to Araújo *et al* (2019) in their work on *crotalaria* juncea (*Crotalária juncea* L., fabaceae) and guandu bean (*Cajanus cajan* (L.) Millsp., fabaceae) crops, the values found in table 3 of the present work are low compared to the values found in their research, which recorded an average of around 20,400 nodules per plant in soil without fertilization.

The same occurs in the results of the study by Ferreira *at al* (2008). In their study of nodulation and agronomic performance of common bean in an agro-ecological production system, the values found in the present work are lower compared to their results, which rose to 60 in the same species under discussion.

Thus, the number of nodules per plant in the *Phaseolus vulgaris* L species found in this work is not in line with the results found by the authors Oliveira (2016), and Oliveira & Sbardelotto (2011), where they found values of 22.4 and 15.0.

Araújo *et al.* (2019), in their work on bacterial nodulation in *crotalaria-juncea* (*Crotalaria juncea* L., fabaceae) and pigeonpea (*Cajanus cajan* (L.) Millsp., fabaceae) crops, provided a higher average than this study due to the delay in sowing by the *Cajanus cajan* (L.) species, which increased its number of nodules per plant to 33,333 compared to the average found in Table 3.

4.2 Nodule weight in grams per plant

Table 8: Weight of nodules in grams per plant

Treatments	Mean±Standard Error
Crotalaria	0,063±3,610[c]
Boer beans	0,045±0,005[c]
Nhemba beans	1,686±0,212[ab]
Kidney Beans	0,118±0,030[c]
Soy	2,688±0,312[a]
CV (%)	18,435
Overall average	1,0

Means followed by the same letters in the columns do not differ when *Tukey's* test is applied at the 5% error level.

According to the analysis of variance at the 5% probability level submitted to the *Tukey* test, the averages relating to the weight of nodules per plant showed significant differences, with the soybean crop being superior in terms of PNP with an average of 2.688±0.312. The *Cajanus cajan* (L) species had the lowest average weight of nodules per plant with 0.045±0.005 g/p

With regard to the weight of nodules per plant, the species *Vigna unguiculata (L), Phaseolus vulgaris* (L) and *Crotalária juncea* L were in the middle with averages of 1.686±0.212, 0.063±3.610 and 0.118±0.030 g/p as shown in table 8 above.

In terms of the weight of nodules per plant, treatment 3 (soybean) had the highest weight of nodules per plant with an average of 2.686g, which disagrees with the results obtained by Santos (2018). In his work on the co-inoculation of

Azospirillum brasilense and *Bradyrhizobium japonicum* in soybeans as a strategy for increasing productivity and reducing nitrogen use, he showed low values with averages of 0.92g compared to the result of treatment 3 (soybeans).

The results of Infante (2008). In his study analyzing the formation of root nodules and nitrogen fixation in 6 varieties of nhemba bean (*Vigna unguiculata* (L)) under water stress, he reports that the average weight of nodules per plant reached 0.85 grams per plant in the Namarua variety, a result considered low compared to the present study, which reached an average of 1.686 grams in the same parameter.

The results obtained in this study for the *Phaseolus vulgaris* (L) species in Table 8 are not in agreement with Araújo *et al.* (2007). In their study of biological N2 fixation in bean plants, the dosages of inoculant and chemical seed treatment compared to nitrogen fertilization recorded the lowest average of around 0.03g of nodule weight per plant in their non-inoculated study, when compared to the results of the present study, which found an average of 0.118g.

Studies carried out by Oliveira *et al.* (2009). Specifically, biological nitrogen fixation in pigeonpea (*Cajanus cajan* (L.) Millsp cv. BRS Mandarim) inoculated with strains of Bradyrhizobium spp. in the presence or absence of fungicide treatment. They report that the weight of nodules per plant can reach 0.8g per plant as an indicator of good nitrogen fixation in this crop, which for the present study was a low average of 0.045g per plant without inoculation of the seeds, probably because in the previous study the inoculated treatments the bacteria proliferated to the point of influencing the increase in the weight of nodules per plant.

Results obtained by Brocardo (2013). In his study characterizing and evaluating the symbiotic efficiency of diazotrophs isolated from bracatinga, he states that the weight of nodules per plant, when it is greater, is an indicator of good nitrogen fixation. In his study, the weight of nodules was 0.239g, whereas in the present study it was 0.063g.

4.3 N2 Before sowing and N2 after harvesting

Table 9: N2 before sowing and N2 after harvest in Kg/ha.

	Nitrogen Before Sowing (kg/ha)	Nitrogen After Harvest (kg/ha)
TREATMENTS	MeanStandard Error	MeanStandard Error
SOYBEANS	42.6 ± 14.449^c	450 ± 6.124^a
NHEMBA BEANS	87.2 ± 1.924^a	417 ± 12.042^a
CROTALARIA	56.4 ± 4.037^b	423 ± 17.176^a
VULGAR BEANS	$47.4\pm2.408b^c$	292.4 ± 112.611^b
BOER BEANS	35.8 ± 3.768^c	409.0 ± 12.449^a
CV (%)	13.409	11.645
GENERAL MEDIA	53.88	398.28
Pr	0.00000007	0.00036

Means followed by the same letters in the columns do not differ when Tukey's test is applied at the 5% error level.

According to the analysis of variance at the 5% probability level submitted to *Tukey*'s test, the averages relating to N2 fixation before sowing showed significant differences, with the plots where nhemba beans were allocated being superior in terms of N_2 before sowing in kg/ha with an average of 87.2±1.924.

It was also observed that the plots where pigeonpea was allocated were lower in terms of N fixation$_2$ before sowing in kg/ha with an average of 35.8±3.768.

The plots where the soybean, common bean and crotalaria crops would be planted, in terms of N_2 before sowing in kg/ha, are in the middle range of 42.6±14.449, 47.4±2.408 and 56.4±4.037, respectively.

In relation to N_2 after harvest, according to the analysis of variance at the 5% probability level, subjected to the *Tukey* test, the averages relating to nitrogen after harvest showed significant differences, with the soybean crop coming out on top with around 450±6.124 N/kg/ha and the lowest average being seen in the common bean crop with around 292.4±112.611Kg/ha fixed in the soil as illustrated in table 9 above.

Simione *et al.*, (2014). In their work on the potential of intercropping grasses and forage legumes in tropical pastures, they agree with the results obtained in the present study, where they state that soybeans have the capacity to fix between 17 and 450 kg/ha of N in the soil$_2$.

It was also observed that the treatment with the lowest average in relation to the variable N_2 after harvest in kg/ha, was the common bean crop with an average of 292.4±112.611. This work disagrees with the results obtained by Brito (2013) in his study on the Initiation of Nodulation in Bean Cultivars. Results reported BNF contributions in field conditions ranging from 25 to 125 kg/N/ha.

The results relating to N2 after harvest, according to the analysis of variance at the 5% probability level submitted to the *Tukey* test, the averages relating to nitrogen after harvest showed significant differences, so the plots where the cowpea, nhemba bean and crotalaria crops were allocated were in the middle in terms of N2/kg/ha with averages of 409.0±12.449, 417±12.042 and 423±17.176.

According to the study carried out by Filho (2015), in his work on the cultivation of sugar sorghum in succession to crotalaria and the application of nitrogen, the results obtained in the present work disagree, showing a range of N2 fixation from

150 to 450 kg/ha of N2 fixed, with these data disagreeing with those of the present work, which show an average of 423±17,176 kg/N/ha.

In relation to N2 fixation after harvest, the plots with pigeonpea showed low averages of the intermediate ones with an average of 409.0±12.449 N/kg/ha, results that disagree with Filho (2015), who found that pigeonpea is in the range of 37 to 280kg/ha, as well as those of Simione (2014). His N2 fixation results for the pigeonpea crop are lower compared to those of this study, which are in the range of 7-235kg/ha of N_2 .

For the plots allocated to nhemba beans, the averages relating to N_2 after harvest showed significant differences between the treatments, at the 5% probability of error level, with an average of 417±12.042 N/kg/ha. The results obtained in this study are in disagreement with those of Simione (2014), where in his study of the potential of intercropping grasses and forage legumes in tropical pastures, the results obtained were in the range of 73-240 kg/N/ha. These values are considered to be lower compared to the present study.

4.1.1 Fixation of atmospheric nitrogen in the soil

For this study, the results obtained on the rate at which legumes fix atmospheric N_2 in the soil, it was necessary to obtain laboratory results on N_2 before sowing, this nitrogen that the soil naturally had before sowing the crops studied, which are legumes.

The N2 after sowing was obtained by harvesting the soil when the crops had reached 75% flowering and submitting it to the laboratory. The results for each crop are shown in the table below for a better understanding after subtracting the nitrogen before sowing.

Table 10: Amount of N₂ fixed in the soil in Kg/ha

Legumes	N2 Before sowing kg/ha Mean±Standard Error	N2 after harvest kg/ha Mean±Standard Error	Fixation rate N2 in kg/ha
Soybeans (*Glycine Max*)	42.6±14.449	450±6.124	407.4
Crotalaria (*Crotalaria juncea*)	56.4±4.037	423±17.176	366.6
Nhemba beans (*Vigna unguiculata*)	87.2±1.924	417±12.042	329.8
Cowpea (*Cajanus cajans*)	35.8±3.768	409.0±12.449	373.2
Kidney Beans (*Phaseolus vulgaris*)	47.4±2.408	292.4±112.611	245

The results obtained for the two variables had to be subtracted from the value found for N2 after harvest, in order to obtain the veracity of the objectives, where it was possible to use a simple formula NDC- NAS=NFC (Nitrogen after harvest minus Nitrogen before sowing equals Nitrogen fixed by the crop).

CHAPTER V. CONCLUSIONS AND RECOMMENDATIONS

From the results obtained in this work, it can be concluded that:

The Soybean crop came out on top in terms of the number of nodules per plant with around 24.12±4.820 and the lowest average was seen in the Vulgar Bean crop with around 11.40^1.738.

The Soybean crop was superior in terms of nodule weight per plant with around 2.688±0.312g/p and the lowest average was seen in the cowpea crop with around 0.045±0.005g/p

The nhemba bean crop came out on top in terms of pre-sowing nitrogen with around 87.2±1.924 kg/ha and the lowest average was seen in the pigeonpea crop with around 35.8±3.768 kg/ha.

The Soybean crop came out on top in terms of post-harvest Nitrogen with around 450±6,124 kg/ha and the lowest average was seen in the common bean crop with around 292.4±112,611kg/ha.

The results obtained in this study provide sufficient evidence to accept the alternative hypothesis that at least one of the five legumes will show significant differences when it comes to fixing atmospheric nitrogen in the soil.

5.1. Recommendations

This work recommends:

5.1.1 CIAM-IIAM-Mapupulo:

That more similar studies be carried out to better guarantee and verify the veracity of the results for the Soybean (*Glycine Max*) 450±6.124, *Crotalaria (Crotalaria juncea)* 423±17.176, Nhemba Bean (*Vigna unguiculata* L) 417±12.042, Boer

Bean (*Cajanus cajans*) 409.0±12.449 and Vulgar Bean (*Phaseolus vulgaris*) 292.4±112.611 species.

That studies be carried out in the legume sector to evaluate the fixation of atmospheric nitrogen in the soil under field conditions and for family producers.

5.1.2 To the producers

> That they choose to grow soybeans if the aim is to improve the texture and structure of the soil, and even to obtain higher yields from crops that are intercropped.

> Use soybeans, nhemba beans or boer beans for intercropping with grasses and even crop rotation.

5.1.3 To agricultural extension workers

The results of this study should be disseminated in order to help family growers choose the right crop for intercropping.

For the recovery of degraded soils, green manure, controlled fallow in compositions with low fertility, opt for *crotalaria juncea.*

CHAPTER VI. BIBLIOGRAPHICAL REFERENCES

Arruda, N. (2012) Evaluation of the structure and physiological potential of crotalaria seeds using image analysis resources.

Araújo, V.A. (2015) Physical, physiological and analogical characterization of *crotalaria juncia* L. seeds harvested at different stages of maturity.

Alfredo, A. J. (2013). Evaluation of the Performance of Nhemba Bean Genotypes (*Vigna unguiculata* (L.) Walp.) and Stability of Grain Yield in Southern Mozambique. Maputo.

Anjos, A. (2005). Design of experiments I. Available at https:///docs.ufpr.br/aanjos/CE213/apostila.pdf

Araújo, A. L. S., Pinheiro, A. R., Silva, A. M., Araújo, A. S., Costa, A. L. M., Barros, P. R. (2019). Nodulation of bacteria in cultivars of sun hemp (*Crotalaria juncea* L., fabaceae) and pigeon pea (*Cajanus cajan (*L.) Millsp., fabaceae) with different sources of organic matter. Journal of the State University of Alagoas/UNEAL. Available atfile: ///C:/Users/Downloads/160-Texto%20do%20artigo-329-1 -10-20191223%20 (1).pdf

Araújo, S. A., Pinheiro, A. R., Silva, M. A., Araújo, S. A., Costa, A. L. M., Barros, P. R. (2019). Bacterial nodulation in *Crotalaria juncea* (*Crotalária juncea* L., fabaceae) and guandu bean (*Cajanus cajan (*L.) Millsp., fabaceae) crops with different sources of organic matter. Revista Ambiental. E-ISSN 2318-454X, Year 11, Vol. 11(3).

Araújo, C. K. (2014). Evaluation of promising pod bean (*Phaseolus vulgaris* L.) lines in cambuci-r for study of cultivation and use value. Brazil. RJ

Araújo, F. F., Carmona, G. F., Tiritan, S. C., & Creste, E. J. (2007). Biological N2 fixation in the bean plant at inoculant dosages and chemical seed treatment compared to nitrogen fertilization. Agronomy course, Faculty of Agricultural Sciences, Universidade do Oeste Paulista. SP. Br

Bohrer, J. R. T. & Hungria, M., (1997). Evaluation of soybean cultivars for biological nitrogen fixation. Embrapa-National Soybean Research Center (CNPSo).

Brito, F. L. (2013). Nodulation initiation in bean cultivars. Federal Rural University of Rio de Janeiro. Br

Brandelero, M. E., Peixote, P. C., & Ralisch, R. (2009). Nodulation of soybean cultivars and its effects on grain yield. Ciências Agrárias, Londrina, v. 30. Available at
http:///www.uel.br/revistas/uel/index.php/semagrarias/article/viewFile/3559/287
3

Brocardo, E. M. C. N. (2013). Characterization and evaluation of the symbiotic efficiency of diazotrophs isolated from bracatinga. CAV/UDESC. Retrieved from http:///www.cav.udesc.br/arquivos/idsubmenu/832/ehrhardtbrocardo_natalia_c._m..pdf

Castro, M. C., A, R. J. B., A, L. D., & R, D. L. (2004). Green manure as a source of nitrogen for eggplant cultivation in an organic system: Federal Rural University of Rio de Janeiro.

Carellos, C. D. (2013). Evaluation of guandu bean cultivars (*Cajanus cajan* (l.) millsp.) for fodder production in the dry season, in são João evangelista-mg. Minas Gerais - Brazil.

Diniz, B. L. M.T. (2006). Cultivation of common beans. Santa Catarina : UFSC.

Oliveira, S. P. A. (2016). Performance of common bean inoculated with rhizobium in response to different cover crops and dissection times. Goiânia, GO - Brazil

Embrapa (2011). Production system. Soybean production technologies central register of Brazil, 1ª edicao. Londrina.

Ferreira, E.P.B., Santos, R.F.,Mata, W.M., Coelho, L.H., Barbosa, L.H.A., & Didonet, A.D. (2008). Nodulation and agronomic performance of common bean in an agro-ecological production system. Uni-Anhanguera University Center.

Ferreira, L. A. (2015). Study of bean cultivars (*Phaseolus coccineus* L.) as potential rootstocks in the cultivation of green beans (*Phaseolus vulgaris* L.). Brazil. SP

Filho, G. G. H. (2015). Cultivation of Sugar Sorghum, in Succession to Crotalaria and Application of Nitrogen, in the South of Roraima.UERR edicoes.BR

Hungria, M., Campo, J. R. & Mendes, C. I. (2001). Nitrogen fixation in soybean cultivation/Londrina: Embrapa soja. Retrieved from https:///ainfo.cnptia.embrapa.br/digital/bitstream/CNPSO/18515/1/circTec35.pdf

Júnior, F. I. P., & R eis, M. V. (2008). Some Limitations to Biological Nitrogen Fixation in Leguminosae. Seropédica/RJ Brazil.

Lamônica, R. K. (2008). Benefits of crotalaria on the nutrition and growth of mango, soursop and nem and on changes in soil characteristics in agroforestry systems. RJ. Brasil.

Martin, N. T. (2016). Soybean cultivation (*Glycine max* L). Federal University of Santa Maria.Br

Ministry of State Administration (2005). *Profile of the Montepuez District.* Ministry of State Administration. 2005. Available at: http:///www.govenet.gov.mz/Acesso on February 10, 2015.

Melo, R. S. & Zilli, E. J. (2009). Biological nitrogen fixation in cowpea cultivars recommended for the state of Roraima. EMBRAPA. BR. Available at https:///www.scielo.br/pdf/pab/v44n9/v44n9a16.pdf

Mendes, F. T. (2019). Productivity of soybean cultivars as a function of plant density variation. Brazil. Rio verde-go.

Macedo, L. F. (2010). Estimation of N2 fixation through sap N composition in Anadenathera falcata. São Paulo. Available at http:///arquivos.ambiente.sp.gov.br/pgibt/2013/09/Fernanda_Lopes_de_Macedo _MS.pd f

Matoso, G. C. S. (2013). Contribution of size-stratified nodules to biological nitrogen fixation in bean. Sc. Br

Oliveira, C. R., & Sbardelotto, M. J. (2011). Nodulation in different bean varieties inoculated with *Rhizobium tropici*. Assis Gurgacz College - FAG. Br

Oppewal, J., & Cruz, A. (2017). Analysis of the Boer bean value chain in Mozambique: public policies and action plan. Maputo. IGC. Available at https:///www.theigc.org/wp-content/uploads/2017/12/Relatorio-F.Boer-FINAL2.pdf

Oliveira, D. M. A , Torres, F. F., Mendonça, G. G., Capristo, P. D., Lima, C. D. (2018). Soybean cultivation with low environmental impact (bia). CONTECC. Maceió-AL, Bras

Oliveira, A. P. P., Schaer, A. B., Ascencio, F., Godoy, R., & Tsai, M. S. (2009). Biological nitrogen fixation in guandu (*Cajanus cajan* (L.) Millsp cv. BRS

Mandarim) inoculated with *Bradyrhizobium spp.* strains in the presence or absence of fungicide treatment. Br

Padovani, C. R. (2014). Design of experiments. São Paulo State University

Pereira, G. R,. Albuquerque, W. A., Souza, O. R., Silva, D. A., Santos, A. P. J., Barros, S. A., Medeiros, Q. V. P. (2011). Soil management systems: soybean [*Glycine max* (l.)] intercropped with *Brachiaria decumbens* (stapf). Goiânia, v.41.Br

Pinto, V. J. (2016). Physical, chemical, nutritional and technological properties of beans (*Phaseolus vulgaris* L.) of different color groups. Federal University of Goias. Brazil.

Rufini, M. (2013). Efficiency of Bradyrhium ssp. strains in symbiosis with guando cvs-fava-larga under different conditions. Br Embrapa.

Silva, L. R., Beline, C. M., Caramelo, A. D., Galdiano, F. R., Moreira, Q. M. W. (2014). Efficiency of the symbiotic association of bradyrhizobium for the growth of crotalaria (*Crotalária juncea*). SP. Brazil UNIFAFIBE University Center. Available at http:///www.unifafibe.com.br/revistasonline/arquivos/revistafafibeonline/sumario/33/18 122014194453.pdf

Silveira, p. S (2010*),* Sowing time and plant density in peanut cultivars in some sands of the State of Paraíba 4 ed.

Souza, M. G. L. (2016). Optimization of biological nitrogen fixation in soybeans as a function of reinoculation in direct underplanting cover. Brazil Embrapa Londrina. Available athttp:///www.agrisus.org.br/arquivos/relatorioparcialPA1505_segundo.pdf

Silva, K. S. (2018). Nitrogen fixation and transfer in intercropped and single crops of grasses and legumes. Brazil Federal Rural University of Pernambuco.

Silva, B. M. E., Guimarães, J. R. S., Neves, R. C. L., Silva, A. J. T. (2012). Development and production of crotalaria fertilized with reactive natural phosphate in cerrado latosol. Enciclopédia biosfera, Centro Científico Conhecer - Goiânia, v.8. Br.

Souza, G. F., Frigeri, T., Moreira, A., & Godoy, R. (2007). Guandu seed production. São Carlos, SP.

Souza, O. J., Cardoso, N. P., Chaves, C. F. L. (2013). Quantification of nodules of native rhizobia isolated in cowpea [*Vigna unguiculata* (l) walp.] BR. Available at http:///www.eventosufrpe.com.br/2013/cd/resumos/R0417-1.pdf

Seffetrt, N.F &Thiago,L.R.L (1983) Leguminera; forage crop for protein production campo grande; EMBRAPA

Santos, D. A. (2013). Feasibility of inoculating soybean seeds with commercial products based on bradyrhizobium japonicum before sowing. Federal University of Paraná. Br. Available at https:///acervodigital.ufpr.br/bitstream/handle/1884/35143/TCC%20Adrianeapro vado.p df?sequence=1&isAllowed=y

Simione, T. A., Gomes, J. F., Botine, A. L., & Mousquer, J. C. (2014).Potential of intercropping grasses and forage legumes in tropical pastures. Brazil

Santos, K. M. M. (2018). Co-inoculation of azospirillum brasilense and bradyrhizobium japonicum in soybeans as a strategy for increasing productivity and reducing nitrogen use. Evangelical College of Goianésia. Brazil

Vieira, F. R. (2017). Nitrogen cycle in agricultural systems. 1ª Edition. Br. Embrapa.

CONTENTS

Printed by Books on Demand GmbH, Norderstedt / Germany